QUANTUM PHYSICS FOR BEGINNERS

A Short and Smart Introduction to The Quantum Physics Theories made Easy

By
EUGENE MAXWELL

© Copyright 2020

All rights reserved.

This report is towards furnishing precise and reliable data concerning the point and issue secured. It was conceivable that the manufacturer was not required to do bookkeeping, legally approved, or anything else, competent administrations. If the exhortation is relevant, valid, or qualified, a rehearsed person should be requested during the call.

The Declaration of Principles, which the American Bar Association Committee and the Publishers and Associations Committee have accepted and supported.

It is not appropriate to reproduce, copy, or distribute any portion of this report in either electronic methods or in the community written. Recording this delivery is carefully disclaimed, and the ability of this report is not permitted until the distributor has written a license. All rights. All rights are held.

The data provided in this document is expressed in an honest and predictable way, like any risk, in so far as abstention or anything else, is a singular and articulate duty for the beneficiary peruser to use or mistreat any approaches, procedures, or bearing contained within it. No legal responsibility or blame shall be held against the distributor for any reparation, loss, or money-related misfortune because of the results, whether explicitly or implied.

All copyrights not held by the distributor are asserted by individual authors.

The details herein are solely for educational purposes and are all-inclusive. The data was entered without a contract or acknowledgment of assurance.

The marks used shall be without consent, and the distribution of the mark shall be without the consent or support of the proprietor of the mark. All trademarks and trademarks within this book are just for explanation and are held clearly by the owners, who are not associated with this record.

Table of Content

INTRODUCTION 5

CHAPTER 1: THE DEVELOPMENT OF A THEORY 11
- THE STRANGE QUANTUM PHENOMENA 12
- UNCERTAINTY IS THE KEYWORD 13
- LET'S TALK ABOUT NETWORKS SECURITY 15
- TOWARDS SUPER FAST QUANTUM COMPUTERS 17
- NEW HYPOTHESIS 19
- NEW TYPES OF MATTER 21
- ALBERT EINSTEIN AND THE BIRTH OF QUANTUM PHYSICS ... 28
- THE QUANTUM HYPOTHESIS AND THE STUDY OF LIGHT 31
- QUANTUM THEORY OF LIGHT 35

CHAPTER 2: QUANTUM PHYSICS AND LAW OF ATTRACTION 45
- HOW DOES THE PRINCIPLE OF ATTRACTION APPLY? 53
- QUANTUM NEUROSCIENCE 55
- EXAMPLES OF QUANTUM PHYSICS PRESENT IN OUR DAILY LIFE 62

CHAPTER 3: THE TEQ EXPERIMENT 67

CONCLUSION 71

INTRODUCTION

How many of you think that quantum physics is only for the "insiders"? It sounds like a strange word, and even for physicists who deal with it every day, it can seem not very easy at times. However, quantum physics can also be understood by beginners.

In fact, throughout the universe, there is the presence of both waves and particles. Everything inside the universe, thanks to quantum physics, finds an explanation in these wave concepts.

Of course, as we have already said, there is also the existence of particles and waves in the universe. It doesn't seem very easy to understand. Still, it is an experimental reality created through a method well known with quantum mechanics, a science of physics dedicated to studying the behavior of the particles of which matter is composed and to the study of energy.

This science allows us to give a modern interpretation of the study and description of supermassive objects such as stars and galaxies and to be able to understand and investigate cosmological phenomena, such as the Big Bang.

The word "quantum mechanics" was invented by Max Born in 1924.

The general physics community has accepted quantum mechanics because it tries to predict systems' physical behavior, even those in which the principles of Newtonian mechanics fail.

Always speaking in terms of quantum mechanics, we can observe how general relativity itself is not limited to the definition of structures at atomic or lower levels, with very low or very high energy or at lower temperatures.

The theory of quantum mechanics has been efficient and realistic thanks to a century of experimentation and evaluation.

As we have already mentioned, the origins of quantum mechanics date back to the early 1800s, but the real beginning dates back to 1900, thanks to

Albert Einstein and Niels Bohr developed with their works what is now identified as the "old quantum theory."

However, it will be necessary to wait until 1924 to get a more detailed picture thanks to Louis de Broglie and his theory of the wave of matter, thanks to which we understood what quantum mechanics really meant.

As for what is now called "modern quantum mechanics" or "modern physics," we owe its birth to the work of scientists Max Born, Paul Dirac, Werner Heisenberg, Wolfgang Pauli, and Erwin Schrödinger.

Subsequently, Julian Schwinger, Sin-ItiroTomonaga, and Richard Feynman in 1947 developed quantum electrodynamics. Murray Gell-Mann gave birth to quantum chromodynamics, which created a model representing light as a particle that cannot describe the interference that creates bands' colors on a bubble.

However, this phenomenon can be explained by a model that portrays it as a wave.

Early researchers gave different explanations in their interpretations of the existence of what we now call electromagnetic radiation.

Some have claimed that light and other wavelengths of electromagnetic radiation are made up of particulate matter, while others have claimed that electromagnetic radiation is a wave phenomenon.

In the field of classical physics, these theories are incompatible with each other.

So early on, scientists understood that no theory alone could describe electromagnetic radiation.

In quantum mechanics, microscopic objects' rules and behavior are very different from those we live in our daily experience, thus generating certain disbelief.

It is now widely believed that most classical physics consists of special quantum physics and/or relativity theory.

Dirac introduced the theory of relativity to quantum mechanics to adequately address events that occur at a large fraction of light speed.

However, classical physics also analyzes the effects of gravity are, and with the relativized quantum theory, no one has yet been able to insert gravity into a unified theory.

Quantum mechanics is the branch of physics that examines the submicroscopic world. In this context, objects are smaller than we can directly observe with our senses and are usually observed with instruments. Parts of quantum mechanics appear as strange and bizarre as parts of relativity. But science and experiments have proven the validity of quantum mechanics just as they did for relativity.

Today, we accept that matter is made up of atoms, the smallest unit of an element, which combines to form molecules, the smallest unit of a compound. While, for example, we cannot see the various water molecules in a stream, we are aware that this is due to the small and numerous molecules in that stream. When we introduce atoms, we usually say that electrons orbit atoms around a small nucleus in discrete shells made up of smaller particles known as protons and neutrons. We also know that electricity comes in small units almost entirely carried by electrons and protons. We do not notice individual charges in the current through the bulb because the charges are so small and in such large quantities that we feel them directly in macroscopic situations.

Quantum physics is the absinthe of the 1900s, having successfully explained phenomena such as radioactivity and antimatter. In contrast, no other theory has been able to describe light and particles' behavior on smaller scales.

Quantum objects can exist simultaneously in different states and places, requiring a mastery of statistics. The theory has been overshadowed by the confusion and paradoxes seen in it questioning the concepts of objective truth, a problem that many physicists, including Albert Einstein himself, have struggled to accept.

Today, physicists are trying to harness bizarre quantum properties to advance technology by combining quantum physics and general relativity with quantum gravity into a unitary theory.

CHAPTER 1: THE DEVELOPMENT OF A THEORY

During the early 1900s, the quantum theory began to develop when classical theories could not describe any results. Previous theories allowed atoms to vibrate at any frequency, leading to erroneous projections that infinite amounts of energy could be radiated, a problem known as a UV disaster.

In 1900, Max Planck solved this problem by assuming that atoms could only vibrate at certain frequencies or quantities. In 1905, Einstein solved the photoelectric effect's mystery, with electrons of some energies being emitted by light falling on the metal. The established theory of light as waves did not elucidate the effect. However, Einstein proposed a beautiful solution, implying that light was present in discrete energy packets called photons, the brain wave that earned him the Nobel Prize in Physics in 1921.

THE STRANGE QUANTUM PHENOMENA

Depending on the experiment, indeed, the camellike potential of light to act as a particle or wave has long been the subject of discord among scientists. Thanks to the Danish physicist Niels Bohr, this wave-particle duality was explained because he was able to rely only on his own observations without preconceptions. In his "Copenhagen interpretation," Bohr explained that measurement itself is already capable of influencing what we see.

A recent experiment challenged this scenario by detecting both wave and particle data simultaneously. The work suggests that light is quantized due to its interaction with matter.

Furthermore, at least six quantum physics theories propose far broader ideas of a universe based on measurement to solve the problem of measurement itself. Such theories suggest that quantum objects show that they have many different characteristics because they exist in an infinite number of parallel universes.

UNCERTAINTY IS THE KEYWORD

Heisenberg's uncertainty principle has explained this wave-particle duality for over 70 years. This theory was proposed by Werner Heisenberg in 1927 and stated that one could never know either the position or the momentum of one quantum object, invariably measuring one displacement of the other.

In the 1920s and 1930s, Boher defeated Einstein using this theory. Later, it was claimed that the phenomenon known as interconnection is the underlying cause of the duality found in the experiments.

Thus we begin to talk about the concept of "entanglement" (which Einstein calls "remote speaky behavior"), i.e., that in the quantum world, objects, if they interacted or were created in the same process, are interconnected and therefore, any change occurs one always affects the other, no matter how far apart they are.

This concept can explain why objects have mass and include super conductance. It allows you to start talking about things that look like science fiction, like "teleporting" of particles over large distances, as everyone decides on a benchmark. In fact, the first quantum teleportation occurred in 1998, and scientists eventually trapped more and more particles, various forms of particles, and large particles.

LET'S TALK ABOUT NETWORKS SECURITY

Considering the concept of entanglement expressed above, this could also provide an almost indestructible communication tool. In fact, thanks to it, quantum cryptographers can send "keys" with quantum particles to decrypt the encrypted details, and any attempt to intercept the particles would change their number, which can then be detected.

In April 2004, some Austrian financial institutions made the first crypto money transfer. In June, the first encrypted computer network was created with more than two nodes spanning ten kilometers from Cambridge, Massachusetts, USA.

But being able to keep quantum particles entangled is a tricky business. Researchers focus on optimizing the particle transit path and signal. In the UK, scientists recently managed to send encrypted photons through a 100-kilometer fiber optic cable using a sensitive photon detector. Simultaneously, in the US, in an attempt to create someday a quantum connection between the US cities of Washington, DC, and New York, the researchers built a scheme to connect successive clouds of atoms.

TOWARDS SUPER FAST QUANTUM COMPUTERS

The development of quantum computers is another ambitious long-term goal. Since taking advantage of the fact that quantum particles can exist simultaneously in many ways, they would be able to perform multiple computations at the same time, enabling billions of times more computations than traditional computers in seconds.

However, it is a complex situation as particles must remain isolated for computation long enough to preserve their multi-state existence. However, some developments have occurred in this area. Indeed, in 2001, physicists were able to avoid light in their way and overcome a real challenge, while in 2002, the first quantum measurement with a calcium ion was performed. In 2003 a trio of electrons was woven into a semiconductor, and with light, the first quantum logic gate was introduced - the brain behind quantum computers. In October 2004, a string of cesium atoms gave birth to the first component of quantum memory.

But the material particles interact rapidly with each other (only billionths of a second) so that their quantum state is maintained for concise periods. On the other hand, photons hold their states about a million times longer because they are less likely to communicate. However, they are often difficult to memorize as they practically fly at the speed of light.

NEW HYPOTHESIS

While quantum theory describes phenomena that occur on a microscopic scale using three of the four fundamental forces of nature, its weak point is discovered when it comes to the force of gravity. Indeed, this force operates on a much larger scale, and quantum theory has not described it until now.

To try to bridge this knowledge gap, several unusual hypotheses have been suggested, many of which claim that space-time tissues boil with spontaneous quantum fluctuations - giving rise to sprays of "foam" that could represent wormholes and infinite black holes.

Such foam is thought to have filled the universe during the big bang so that things like stars and galaxies could take shape later.

The most famous quantum gravity theory states that particles and forces arise only 10-35 meters from small bands or strings' vibrations and that space and time on a smaller scale emerge from abstractions known as "spin networks."

A new theory called "doubly special relativity" modifies Einstein's concept of a celestial invariant - the speed of light - and introduces another on an incredibly small scale. And gravity, entropy, and dark energy are used to explain this new theory, and physicists are now designing observations and experiments to test competitive hypothesis.

NEW TYPES OF MATTER

Some researchers think that classical physics should not be taken into account, for example, where the attraction force of the weak weight has prevailed over other forces (in fact, the effect of gravity on neutrons has recently been measured), in fact, it is generally believed that quantum physics has less effect than molecules on light and particles. But if macroscopic objects don't get entangled, they could follow quantum laws.

Of course, taking advantage of atomic or quantum photon troops offers fantastic technical promise. New types of matter called Bose-Einstein and fermionic condensates have been created in atoms that cool to near absolute zero. These have been used to produce laser beams composed of atoms that etch complex patterns on surfaces and lead to superconductors a room temperature one day.

All these expectations indicate that it is potentially the most effective scientific solution for years to come; however, one must be careful.

BUT WHAT EXACTLY IS QUANTUM PHYSICS

It is physics that describes how things work in the universe: the most important explanation we have about the nature of the objects that make up matter in the universe and the forces through which they communicate, so it is the basis behind how atoms work and how they work of chemistry and biology. You, me, and any other object are all in turmoil on the quantum theme, at least somewhere. Suppose you want to understand how electrons travel through a chip, how photons from lights turn a solar panel into an electric current or improve into a laser, or how sunburns happen. You need quantum mechanics.

The complexity and, therefore, what physicists like start here. First, it must be said that there is no specific quantum theory. In the 1920s, as mentioned in the previous pages, Niels Bohr, Werner Heisenberg, Erwin Schrödinger, and others formulated for the first-time quantum mechanics or the fundamental mathematical structure underlying everything that has been theorized. This allows you to answer basic questions, such as understanding how a single particle, or a group of multiple particles changes its position or momentum over time.

However, if we are to understand how things work in the physical universe, quantum mechanics must be combined with some aspects of physics in determining what is considered to be quantum field theory, and this is done primarily using Albert Einstein's special principle of relativity. , which describes events as the rapid passing of objects.

Three distinct theories of quantum fields cover three of the four fundamental forces that matter is concerned with: electromagnetism, which describes the bonding of atoms; strong power of the nucleus, which describes the structure of the nucleus in the center of the atom; and low nuclear power, which explains why some atoms are prone to radioactive decay.

For about five decades, all three hypotheses have been forced together to express what is termed particle physics. While this theory is partially held together by concepts that do not fully tie together, it was the most accurate representation of the fundamental function of the subject ever invented. In 2012, a climax was reached when, based on quantum field theories of 1964, the Higgs boson was detected, a particle that provides mass for all other basic particles.

Conventional quantum field theories work well for understanding the effects of experiments with energy-rich particle accelerators, such as the CERN Massive Hadron Collider, with Higgs, whose sample is as short as possible. At the same time, it gets much more complicated if you're trying to learn. They work in much less abstract circumstances, such as how electrons travel or don't pass through a rigid substance to create a magnet, insulator, or semiconductor.

Millions and trillions of interactions in these light conditions result in the development of "strong field concepts" that overlook a few particular details. The problem is that many important questions in solid-state physics remain unanswered when trying to develop these theories, such as why some low-temperature materials are superconductors with no electrical resistance while they are not if we are at room temperature?

For all these practical problems, no adequate answers have yet been found. The way things work in the real world is very different from the explanation that quantum mechanics, at the simplest level, suggests about how matter works. Quantum particles can be like electrons, scattered in one point, or they can be waves, scattered in space or at the same time in several places and seem to depend on how we want to measure them and, before calculating them, they seem to have no defined properties, leading to a deep mystery also about the presence of real reality.

This lack of complete understanding leads to apparent paradoxes, such as Schrödinger's Cat Story. Due to an inexplicable quantum mechanism (a state known as quantum superposition), he is left both alive and dead. But not halfway. Quantum particles also appear to have an immediate effect on each other, particularly if they are very far apart. In a word coined by Einstein (a true enemy of quantum theory), a true Bamboo phenomenon is called "spectral and boundary behavior" ', which is completely unknown to humanity and forms the basis of modern technologies such as quantum cryptography, ultra-secure, and ultra-powerful quantum computing.

But nobody knows what all this means. Some think we have to accept that quantum physics explains the natural world to such a degree that it is difficult to match our understanding of the broader "classical" setting. Others think there may be a better and more rational explanation that we have yet to discover.

There are still "many elephants in the house in all these theories," in fact, there is, for example, a fourth fundamental force that has not yet been able to explain the quantum theory. Gravity is the realm of Einstein's theory of general relativity, a concept that is obviously non-quantum and doesn't even contain particles. For decades, enormous efforts have been made to bring gravity under the quantum umbrella, which would have elucidated all the fundamental phenomena within "a theory of everything," but unfortunately failed.

Meanwhile, cosmological calculations indicate that over 95% of the universe is made up of invisible and dark energy, things that we still don't understand very well in the standard model today, and uncertainty is the essence of the role of quantum mechanics as part of the chaotic functioning of life that remains uncertain. The world is at the quantum level, so it is an open question whether quantum mechanics is the last term in the world.

ALBERT EINSTEIN AND THE BIRTH OF QUANTUM PHYSICS

The critical developments of quantum physics, however, are due to the research of Albert Einstein of 1905-1917, who confirmed the existence of atoms, discrete localized particles of matter, through his study of Brownian motion, but also particles of discrete electricity, as a quantum of light was as localized as matter. In 1917 Einstein defined the irreducible case in the universe of how matter and radiation interact.

Einstein, he thought faster than he believed. In addition to his brilliant theories on relativity, he was one of the most influential pioneers of quantum mechanics. His three 1905 papers on relativity, Brown's motion, and the quantum of light quantify the radiation field, whereas, for example, Planck only quantified energy.

But Einstein never earns any credit for his achievements. Various significant explanations lead historians of quantum theory, first Planck's quantum of motion, then Einstein's articles of 1905 and his

1909 study of the duality of the wave part, to Niels Bohr's classical quantum theory of the atom in 1913, which was actually opposed to Einstein's quantum theory of light until the mid-1920s. Today, Bohr's "quantum leap" between an electron's stationary states is known as the "photon" of energy.

In addition to quantified energy and understanding the interchangeability of radiation and matter, $E = mc^2$, Einstein was also the first physicist to look at various fundamental aspects of quantum mechanics, such as non-locality, simultaneous action at a distance (1905), and the duality of particles (1909).

Unfortunately, Einstein could never have included any of his quantum theories because they disagreed with his fundamental idea that reality is best defined in continuous field theory, using differential equations that are functions of 'local' variables, particularly the four-vector space-time of its general relativistic principle. Einstein's definition of "local" reality is one in which his theory of relativity limits "action at a distance" to causal effects that move at and below the speed of light.

Einstein believes that quantum theory is good since its statistical predictions (Einstein's series) don't tell us enough about individual structures. Worse, he believed that interconnected systems' wave functions predicted that properties are more readily correlated with spatial separation, violating his theory of relativity. This is what was behind his famous 1935 EPR paradox, but we can see he was thinking of faster-than-light motion in his early work on quantum theory.

THE QUANTUM HYPOTHESIS AND THE STUDY OF LIGHT

The energy of a body of weight cannot be arbitrarily divided into several parts or arbitrarily small. On the contrary, luminous flux energy from a point source is continuously distributed over an increasing volume (according to the Maxwellian theory of light or, more generally, any wave theory).

The wave theory of light, which deals with continuous spatial functions, has purely optical phenomena and will never be replaced by any theory. In so-called "space," a basic structural gap exists between physicists' theoretical conceptions of gases and other ponderable artifacts and Maxwell's theory of electromagnetic processes. Although we find that the body's state is largely determined by the position and velocity of a huge and final number of atoms and electrons, we use continuous spatial functions to characterize the electromagnetic state in a given region. According to the Maxwellian principle, in all purely electromagnetic phenomena, including light, energy is to be defined. According to current physicists' principles, the energy of a weighted body must be represented as a quantity carried by atoms and electrons.

It should be remembered, however, that optical measurements correspond to time averages rather than instantaneous values. Despite the full experimental evidence of the theory used in diffraction, reflection, refraction, scattering, etc., the theory of light, working with continuous spatial functions, can still lead to contradictions in practices when applied to the phenomenon the emission and transformation of light.

Assuming that the light energy is continuously distributed in space, it is easier to understand the results of black body absorption, fluorescence, cathode rays of ultraviolet light, and other related anomalies concerning emission transformation. Of the light.

The energy of a ray of light that spreads from a point source, in line with this hypothesis, is not continuously distributed across an area but consists of a small amount of energy located on spatial points, which moves without divisions and it can only be formed and absorbed as complete units.

So, to conclude, "the higher the energy density and wavelength of the radiation, the more valid are the statistical concepts we have used"; however, we are completely devoid of low wavelengths and limited radiation densities.

We also assume that the low-density monochromatic radiation (in the valid region of the Vienna radiation formula) behaves as if it were composed of several independent energy sources.

The energy of a light beam that spreads from a point source is not continuously dispersed over an increasing space. It consists of a finite quantity of energy quanta, located at points in space that travel without separation and can be produced and absorbed as complete units.

It is not without difficulty that light is constantly transferred into the space in which it travels while experiencing photoelectric phenomena.

However, based on the concept that incident light is made up of quantum energy, the following should be considered for electrons' ejection by light. The quantum of energy penetrates the body's surface layer, and the energy is transformed into kinetic energy of the electrons, at least in part. The simplest way to imagine this is to move all the energy into a single electron; we will assume that is what happens.

QUANTUM THEORY OF LIGHT

Einstein greatly developed his quantum theory of light in September 1909, during a debate at the Salzburg conference. He offered an important insight into the irreversibility of natural processes and argued that the radiation-material relationship involves fundamental structures that were not reversible. And although incoming spherical radiation waves can be imagined mathematically, they are not feasible. He hypothesizes that many bright and singular quantum points superimposed for wave behavior can shape the continuous electromagnetic field.

Although he could not formulate a mathematical theory that fits both oscillatory and quantum processes - wave and partition diagrams, Einstein argued that they were compatible over a decade before the wave and quantum mechanics. And because gases behave according to statistics, he knows that the relationship between waves and particles, which includes probabilistic behavior, is seen in 1916.

When the light showed interference and diffraction, it almost seemed likely that the light would be called a tube.

Maxwell's ingenious discovery that light can be understood as an electromagnetic phenomenon was the greatest advance in theoretical viewpoint since the advent of oscillation theory. One was used as a basic principle for electric and magnetic fields that do not need mechanical explanation.

This journey leads to the so-called principle of relativity. I want to add one of its results because it makes other changes to simple physical ideas. It indicates that the body's inertial mass decreases by L/c^2 as the object releases L energy. The emission of light reduces the object's inertial mass.

Energy is provided as part of the material mass. Therefore, any consumption or release of energy determines an increase or decrease in the substance's weight under consideration. Energy and mass are similar to heat and mechanical strength.

The principle of relativity has changed our understanding of space. Light is not seen as a representation of an imaginary process's condition, but rather as an autonomous object such as air. This hypothesis shares a particular aspect with the corpuscular hypothesis of light, which transfers the inertial mass from the transmission to the receiving body. The theory of relativity does not alter our understanding of radiation composition; it does not affect the distribution of energy in spaces full of radiation.

However, I suppose we are beginning to develop the utmost importance concerning this issue, which cannot yet be investigated. The following comments are largely my personal opinion, or the result of hypotheses not yet adequately reviewed by others. If I introduce you here despite your uncertainty, the reason is not overwhelming confidence, in my opinion, but rather the hope that one or the other of you can address your problems.

There is also an opposite process in the kinetic theory of molecules at each stage involving only some elementary parts (e.g., molecular collisions). However, this is not the case with the fundamental radiation mechanisms.

According to our prevailing theory, an oscillating ion produces an outward spherical wave. The opposite method is not a basic process. The convergence of spherical waves is undoubtedly mathematically feasible, but many emitting entities are needed to realize it. The basic combustion process is not reversible. In this, I assume, our oscillation theory does not end. In this regard, Newton's theory of light emission tends to contain more validity than the theory of oscillation because the energy given to the particle of light, first of all, is not dispersed in infinite space but remains usable for a period of primary absorption.

See the law governing the production of secondary cathode X-rays. When the rays of the primary cathode hit a platform of P1, X-rays are produced. If these X-rays affect the second P2, the cathode rays are produced again, the speed of which is equal to that of the primary cathode rays.

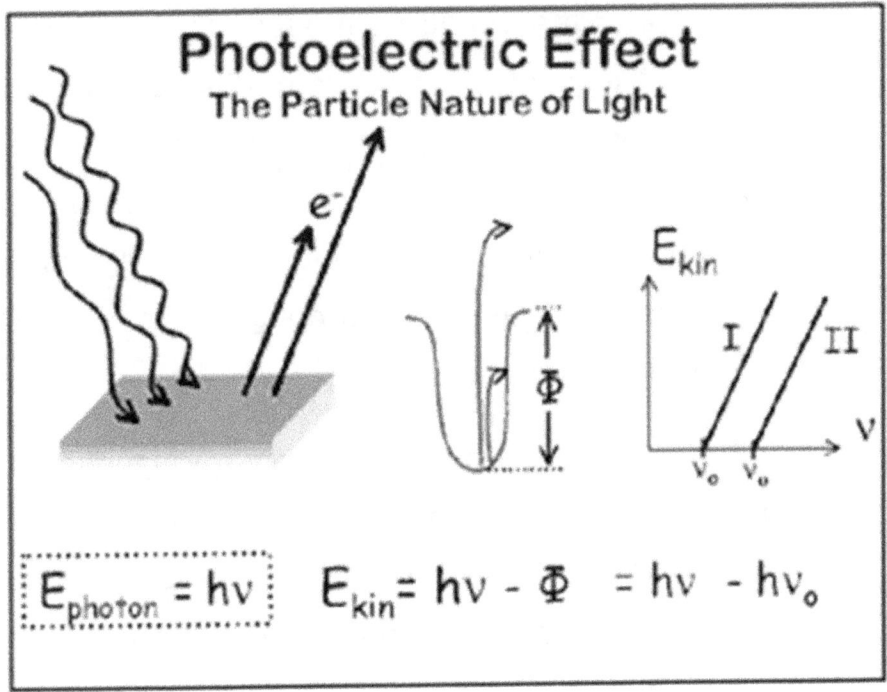

As we now understand, the speed of secondary cathode rays does not only depend on the distance between P1 and P2, and the speed of primary cathode rays but only on the speed of primary cathode rays. That's strictly true, let's say. What would happen if we consider the power of the primary cathode rays or the P1 size they felled to be an independent mechanism for the electron effect of the primary cathode rays?

If otherwise true, since the secondary cathode-ray velocity is independent from the primary cathode-ray velocity, it has to be assumed that an electron that pulses in P1 will induce either the emission of no electron at P2 or a secondary emission of an electron whose speed is equal to that of the original electron in P1. In other words, the primary radiation process appears to proceed in such a way that it distributes the Primary Electron Energy not in any direction through a circular wave that propagates it, as the oscillating theory implies.

Rather, at least a large part of this energy appears at some point to be accessible on P2 or somewhere else. The basic method of emission of radiation appears to be directional. One actually thinks that the production of P1 x-rays and secondary cathode rays processing in P2 are essentially opposite processes.

However, radiation generation seems to be different from what our oscillation theory predicts. This is expressed in the theory of thermal radiation, primarily in the concept on which Planck based his radiation formula ...

The Planck theory applies to the following premise. When it is right that a radioactive resonator can only tolerate multiple HV energy levels, the logical inference is that only in such energy concentrations can light be emitted and absorbed. On the basis of this theory, the light-quantum theory should answer the above issues about light pollution and absorption. The theoretical consequences of this light-quantum principle are valid as far as we learn. That poses the following question. May it not be true that the radiation formula of Planck is right, but another derivation that is not based on such a frightening assumption like Planck's theory can still be identified? Has the light-quantum hypothesis not been substituted by another theory that one does justice to known phenomena? If the elements of theory should be modified, could it not be necessary to keep intact the laws of propagation and only to change the perceptions of the basic emission and absorption process?

To my knowledge, no complex mathematical explanation has been given, which acknowledges both its oscillatory nature and its quantity nature.

This understanding, in both cases, seems the most natural: that at points of uniqueity, the representation of electromagnetic waves of light is reduced, such as the realization of electrostatic fields in the electron theory. This can not be omitted because, in such a case, all the energies of the electromagnetic field can be construed as being concentrated in such singularities, as is the conventional concept of action at a distance. I am pictured in each individual point, surrounded by a field that essentially has the same character as a plane wave, which, with the distance between the singular points, is reduced in intensity. When all these special characteristics are separated by a low gap in terms of their field dimension at a single point, their forces are superimposed to create an oscillating field in its nature that is only slightly different from our existing electromagnetic light theory's oscillating field. For example, such a picture should not be emphasized if it leads to an exact conclusion. I just tried to prove that according to the Planck theorem (oscillational structure and quantum structure), the two structural properties of radiation could not be considered contradictory.

Perhaps the key justification for historians of quantum mechanics (the writings after the Copenhagen definition) to ignore Einstein is the fact that he was the single leading opponent in quantum mechanics in the late 1920s suggested by Niels Bohr, Werner Heisenberg, Max Born, Pascual Jordan, Wolfgang

Pauli, and Paul Dirac. Today, Einstein is better known for his attacks than for his truly outstanding theoretical contributions to quantum theory.

Einstein's main concern is that quantic mechanics is a mathematical theory that forecasts possible results for a wide range of experiments but little in actual circumstances such as the precise period of radioactive decline or spontaneous emission of photons. Einstein himself discovered this "chance" act in human systems in his early papers. However, he was unable to predict real events correctly and concluded that quantum mechanics would remain an unfinished theory.

Quantum mechanics can only describe as many physical amounts as traditional mechanics. According to the indeterminacy principle, only one pair of non-commutable observables (for example, timing or position) can be accurately defined. In this way, classical philosophy is more complete. But instead of merely embracing Einstein's interpretation and terminology, Bohr, Heisenberg, and others took part in the linguistic debate and argued that quantum theory itself is "real."

The second concern of Einstein was that the wave function of an isolated, free particle must expand in time to fill the entire space. All places are equally probable. However, the molecule is always located at a certain location when we detect it. That doesn't mean that the object had a special position before the discovery, but Einstein dedicated himself to "elements of reality," which he thought no one should ask. The location of an atom is one of these components. He asked the question, "Does the particle have a particular location right now before measurement?" Copenhagen's reply was "no," more frequently "do not know." If only one pre-required position exists for the object, its four-dimensional space-time path is set out and decided irrespective of the time. Awareness of the entire path is constant all the time. On the other hand, quantum theory offers alternatives that are necessary if there is to be more than one future possible (with calculable probabilities).

CHAPTER 2: QUANTUM PHYSICS AND LAW OF ATTRACTION

The law of attraction is now a term known to many people, which indicates several ways to improve oneself and improve one's life. The literature contains many interpretations and explains the purpose of the law, but there have been too few attempts to describe the law's physics.

Many have learned the meaning of the law and how it can be applied, but the world is waiting to understand how they can actually benefit from it beyond optimistic thoughts. What are the processes in which it works? Our efforts are focused on enabling people to apply the law more effectively and with less effort. We have found a missing link to how this amazing fundamental theory of magnetism is applied and executed.

Research on quantum mechanics has shown that the act of observing reality produces it. Trying to detect something doesn't allow anything to happen. Similarly, if you don't know anything, in your subjective reality, it doesn't exist. In fact, the so-called placebo effect has shown that positive or negative behaviors have congruent effects.

It becomes clearer and clearer that we help create our reality in the way we think and feel. In other words, our personal and special perception of the truth. We tap into ourselves what we believe is real.

There is only a quantum of energy as a flow of potential spatial positions and movements. This is omnipresent, but it is never found at the same time. Research has shown that the "collapse of the wave function" is none other than consciousness acting manifesting reality.

The wave function involves all possible effects of a given situation, but only one arises when a participating consciousness falls into the physical world. All the scientists look at each other and look at the CERN event's new computer in Switzerland. You expect two protons to crash into each other, and when you do, there is an "episode" on the screen showing the release of all the subatomic particles. They do not realize that wave functions are brought to light by their own observation on the screen. But, since they do not wish to use them in any constructive way, they disappear once again: ~) realizing that each of our realities (body, mind, and spirit) comprises some energy particles whose wave functions have collapsed from our own remarks. Through imagination, inner listening, and a strong sense of our emotions, we will learn to create what we want—experience what we want to manifest in our minds. Likewise, by refusing to give them our attention, we can resist the breaking of possibilities into artifacts, events, and circumstances. Consciousness has to press some of the energy to be real. Until then, it was shrouded in the mystery of probability on the other side of the quantum world. When energy is detected, the "wave function" collapses, making it observable in the real world. Such particles cannot be seen with the naked eye, but only with specialized tools that can show where it was and the speed and position it was detecting.

Since both consciousness and spirit exist on the other side of the quantum world, it makes sense to control our reality, even more so when we are in an inspired or good-humored state. Once we are connected to this original energy source, we have more power to create real and constructive change in our lives. When we linger in this state of grace and respect, we get out of the box. Synchronicity increases, and people want to be around us. A positive wave function will more easily collapse in this mode.

Staying in that condition requires a full emphasis on consciousness. Instant understanding of the divine, the one and all that leads to trust in the universe, leads to the faith needed to reach a higher level. We have to learn how to collapse the positive and useful wave functions while removing our attention from the negative and harmful effects. An individual's thought, speech, and behavior can alter the course of the world. This "observation" phenomenon often applies to other sensory mechanisms. If you hear something from your inner voice, the spirit of reality, or the higher self, or feel something deep in your heart, when you realize it, do the same. You live it. Therefore, the experience is a better word to describe true full life than a wave collapse process, and it does so even better when the experience is multisensory. Sight, hearing, and experience lead to a sensory resonance condition in which all sensory functions are coordinated. In contrast, an autonomous and balanced nervous system leads to a mixture of deep relaxation and extraordinary imagination.

By enhancing their own development and positive spiritual growth, a person draws a positive experience into themselves.

Nature does not rule the world and our lives; they are fundamentally untidy and in turmoil. However, the four fundamental forces that physicists point to materialize reality have recently been redefined as celestial attractors that create order patterns over time. This is the true mystery of the expression of truth. Space is the original force that creates the universe from zero-dimensional points or singularities. It's time to chain the points together into the other attractors.

To use the Law of Attraction, a person must become an attractor.

There are four orders:

1st order POINT Attractor - leads you to get drawn into a certain task or get stuck in a routine by focusing on an idea or obsession for too long.

2nd Order CYCLE Attractor - Allows you to get trapped in abstract thinking or an infinite loop that basically repeats itself over and over.

The 3rd order limit of the attractor is called TORI. This is a step in the right direction as it facilitates a complex energy flow but is somewhat limited by its near periodicity.

The 4th order odd attractor: the Absolute's chaotic action to cancel out all that is deemed to be in disorder.

The emphasis, purpose, and coherence of the action and expression you use determine the attractor you fall into. The best course of action is to coordinate with the 3rd order attractor. In other words, a person could become an infinite toroidal attractor by tapping into what they want rather than taking something that the strange chaotic attractor has to offer. This is the condensed essence of the Law of Attraction and its intellectual and executive secrets.

The first step in learning how to collapse the wave mechanism is to feel what you think and have inside you. In turn, it results from applying the rule to attraction, investigating how to bind to the ideal attractor when paying attention to it in images. Through a two-phase multisensory approach, the user must become a transformative agent by overcoming the disparity and stress between the operator and the system, a polarity that can also be represented as player vs. tool, subjective vs. objective, or real vs. potential, or as the engineers know the real vs. the imaginary. These real and imaginary components are called automatic elements and are the simplest and most inseparable building blocks of existence. We are the operator and the device at the most basic level.

For years, psychologists have known that visualizing a goal in an image is the nature of integrating many scenarios with what the audience actually sees. The example is in white noise to see the inner voice or music. The proof of the model is the incredible ring of truth, which is often perceived in unpredictable forms. Taking such ideas to the quantum level and giving a person a means to imagine their intent, listen to their inner voice, and witness their emotions at the most fundamental truth where life is born is one of our main goals.

The sensory processes that give him information are the only understanding of the relationship between the internal and external worlds. The more sensory mechanisms that are incorporated into the interaction, the more coherent the experience becomes. Each of the sensory mechanisms distinguishes a specific portion of the universal spectrum. Using the multisensory quantum input, you can see, hear, and feel the wave pattern in the most desired result through your will.

HOW DOES THE PRINCIPLE OF ATTRACTION APPLY?

Everything is connected in the universe, and if you remember how it happens with the entanglement principle described above, if we affect one aspect, we will affect another. All we can manage is our emotions. Using our emotions, we touch other objects that we dreamed of communicating with us and draw people, circumstances, and objects to influence us.

The real world is a vortex of energy that constantly bursts into and out of existence. With our reflection, we transform this ever-changing energy into a concrete reality. And we can build our reality with our feelings. Science rejects the quantum physics idea that humans are helpless victims and pushes us to understand that we are actually able to build our lives and our universe.

In Newtonian laws, we are meaningless gears in the universal machine, while in the world of quantum mechanics, "we are the creators of the universe."

Einstein, who is 1905, with his theorem of $E = mc2$, explains how energy and matter are connected - in essence, everything is energy, in motion, flowing and constantly evolving.

Our thoughts influence this ability to shape our reality as creators; we create a shape and shape the universe's energies through our thoughts. The energy of our thoughts will become the energy of our lives.

We have been taught to believe that the external world is truer than the internal reality. Quantum mechanics only suggests the opposite. It means that what happens inside determines what happens outside. This says that our emotions make up our universe.

Everything is conceivable because nothing is established, and everything is possible. If we realize that everything is feasible and focus on what we want to draw in our minds, we will create virtually everything we want.

With all this in mind, we can now fully understand how the illusion works: our special and specific understanding of the truth produces our fact; we carry within ourselves what we see as true. It is the most effective way of manipulating our own. Perceptions - allowing our minds to believe it is.

QUANTUM NEUROSCIENCE

Why all this hype about quantum science?

Quantum science promises leaps and bounds rather than the ant steps that everyday science advances. For example, every 2-3 years, today's science provides us with new computers that double power. On the contrary, quantum science promises computers many trillions of times more powerful than the best computer on the market today.

In other words, when successful, quantum breakthroughs will produce an epochal earthquake in technology that will shape the world as we know it by transforming it in even deeper ways than the internet and smartphones have.

Quantum science has a unique and surprising possibility: quantum phenomena completely break the rules limiting what "classical" (normal) phenomena can achieve.

Two examples of what quantum science suddenly makes possible are quantum superposition and quantum interlocking.

As for the quantum superposition, to understand what it is, we think, for example, of an object like a tennis ball, which can only be in one place at a time in the normal world. Simultaneously, a particle such as an electron in the quantum world can occupy an infinite number of places present in what physicists call multiple superpositions. So one thing sometimes behaves like many different things in the quantum world.

Now let's examine how the tennis analogy is interspersed with the quantum. In the normal world, two tennis balls, one in a locker on a playing field in New York and the other in a locker on a playing field in Chicago, are completely independent of each other, so when in a court, a locker is opened to lay a tennis ball, there is absolutely no other tennis ball in the darkness of the atria 3000 lockers in the court. In the normal world, two tennis balls are completely independent. But in the quantum world, there can be two separate particles, such as photons. The simple act of detecting a photon with one detector instantly forces the other photon to assume a certain state, no matter how far away they are.

This interconnectedness means that multiple distinct entities can occasionally act as a single entity in the quantum universe, regardless of the distinct entities' distance.

It would be equivalent to seeing a tennis ball sitting in a locker in New York while being present in one of the other 3,000 lockers between Chicago and New York.

These "impossible" behaviors make quantum entities suitable for computers, for example. A bit of information stored in normal computers is zero or one, but a stored bit, called Qubit (quantum bit), is all zero and one in a quantum computer at the same time. A simple 8-bit memory store can hold every single number 0-255 ($2 \wedge 8 = 256$); 8-bit memory can store two $\wedge$ 8 = 256 individual numbers all at once! Quantum computers offer a quantum leap in computing power because they can store more information exponentially.

In the example above, an 8-bit memory stores 256 numbers simultaneously between 0 and 255, while an 8-bit memory stores only 1 number in ordinary computers from 0 to 255. Imagine a 24-bit memory ($2 \wedge 24 = 16,777.\ 216$), with only three times the number of qubits in our first memory: a huge amount of 16,777,216 different numbers could be stored at once!

This brings us to the intersection of Quantum Physics and Neurobiology. The human brain is a far more powerful processor than any current computer today - can it gain any of that great strength from using quantum weirdness just like quantum computers do?

Until recently, the physicists' response was a resounding "No."

Quantum phenomena, such as superimpositions, depend on the isolation of surrounding phenomena, particularly heat in the environment, which sets the particles into action, disturbing the hyper-delicate house of quantum cards and forcing a given particle to occupy one of the points A or point B but never both.

Therefore, when scientists research quantum phenomena, they go a long way to isolate the substance from the world they research, usually by lowering their studies' temperature to practically zero.

However, evidence emerges from the field of plant physiology that biological processes based on quantum superposition occur at normal temperatures and raises the possibility that unimaginably strange environments of quantum mechanics could potentially interfere with the daily functioning of other biological systems such as our neurons.

For example, in May 2018, an investigative team from the University of Groningen, consisting of physicist Thomas la Cour Jansen, found evidence that plants and some photosynthetic bacteria achieve nearly 100% efficiency by converting sunlight into energy. Usable by exploiting solar energy absorption allows certain electrons to exist simultaneously in excited and non-enterprising molecules.

Evolution appears to have rejected physicists' hypothesis that the use of quantum effects cannot occur in hot, humid biochemical conditions in its endless search for the most energy-efficient ways of life.

An entirely new science called quantum biology was created by discovering quantum effects in plant biology. In recent years, quantum biologists have uncovered evidence of quantum mechanical properties in detecting magnetic fields in some birds' eyes and activating scent receptors in humans. Vision researchers also found that photoreceptors in the human retina can produce electrical signals by collecting a single amount of light energy.

Was evolution also over-efficient in our brains to produce usable energy or transmit and store information between neurons using quantum effects such as superposition and interlocking?

Neuroscientists are at the beginning of the investigation. Still, I'm excited for one on the new field of quantum neuroscience as it could lead to bottlenecks in our understanding of the brain.

I say this because the history of science tells us that the greatest discoveries almost always come from concepts that sound really weird before a specific breakthrough takes place. Einstein's observation that time and space are essentially the same (general relativity) is an example; Darwin's discovery that humans are born of more primitive life forms is another. Planck, Einstein, and Bohr's first discovery of quantum mechanics is, of course, another.

All of this clearly suggests that most people today believe that the theories behind tomorrow's game turn progress in neuroscience as extremely unorthodox and impossible.

EXAMPLES OF QUANTUM PHYSICS PRESENT IN OUR DAILY LIFE

Quantum physics is perhaps the greatest technological achievement in human civilization's history, but to most people, it seems to be too distant and abstract for the real world.

Any connection between quantum phenomena and everyday life can be difficult to see.

Currently, however, quantum mechanics is all around us. The world, as we know it, is subject to quantum laws. While traditional physics looks very different when quantum physics is applied to many particles, many common and everyday phenomena are induced by quantum effects.

Here are some examples of things in your daily life that you may experience without realizing that they are quantum:

TOASTER:

The red light of a heating element as you toast a slice of bread or a bagel is a well-known sight for most of us. This is also where quantum physics began: it is a problem in quantum physics invented to solve the special color of red and explains why very hot objects glow.

The color of the light emitted by a hot object is an example of a fundamental and universal medical phenomenon. No matter what an object is made of the light spectrum emitted is the same if it survives heating to a given temperature. This kind of universal behavior was discovered in the late 1800s by many brilliant physicists, but none have been able to overcome the problem.

A simple universal approach has been proposed whereby light was independent of composition: take all the light colors that an object could emit and give each an equal share of heat energy in the object. The problem here is that there are far more ways to emit high-frequency light than low-frequency light, which ensures that your toaster can spray x-rays and gamma rays all over the kitchen instead of giving off a nice warm glow. This obviously doesn't happen (thankfully!), So something else has to happen.

Max Planck overcame this problem by proposing the "quantum hypothesis" (which is called the eventual theory) that light can only be released in isolated bits of energy, integer multiples of short constant pulses having the frequency of light. This amount of energy is less than the proportion of heat energy assigned to this frequency for high-frequency radiation, and therefore no radiation is emitted on this frequency. This disrupts the high-frequency light and results in a formulation that matches the spectrum of light observed by hot objects with great accuracy.

Therefore, every time you toast bread, look at where quantum physics begins.

FLUORESCENT LIGHTS:

Incandescent bulbs shed light by having a resistor inside them, which, when passed through by an electric current, overheats and creates a bright white glow that makes them quantum in the same way as a toaster. Get light from another revolutionary quantum process if you have fluorescent bulbs.

As early as the early 1800s, physicists discovered that each element of the periodic table has a specific spectrum: if you get a vapor of hot atoms, they emit light in a small number of distinct frequencies, each with a different pattern. These "spectral lines" were used quickly to classify unknown materials' composition and even discover previously unknown elements: helium; for example, it has been identified in a spectral line previously unknown in sunlight.

Until 1913, no one explained these phenomena, until Niels Bohr took Planck's quantum idea and provided the first quantum model of an atom. Bohr suggested that some special states allow an electron to orbit around an atom's nucleus willingly and that atoms are only absorbed and released by traveling between them. The frequency of absorption or emission of light depends on the states' energy difference and the way Planck implements the model, providing a set of distinct frequencies for each atom.

This was a radical idea, but it worked brilliantly to understand the spectrum of light produced by hydrogen and even the X-rays provided by many elements. Although the modern picture of what is happening in an atom differs considerably from Bohr's model, the central concept is the same: electrons travel within atoms between special states by absorbing and emitting light from certain wavelengths.

This is the central concept behind fluorescent lighting: a little mercury vapor inside a fluorescent bulb (either a long tube or CFL), excited in a plasma. If we gaze at a light (having Mercury released light at wavelengths that typically fell across the visible spectrum so that our eyes assumed the light was white), we can see some distinct colored images of the bulb. While we stare at the light emitted by an incandescent bulb, we notice a continuous rainbow spot. We look at a fluorescent light through an inexpensive diagonal lattice, as you can see in the new glasses.

Therefore, when you use fluorescent lights to illuminate your home or office, you have to be thankful for quantum

CHAPTER 3: THE TEQ EXPERIMENT

More and more physicists nowadays are no longer satisfied with quantum physics textbook concepts derived from Bohr's understanding and other interpretations of quantum theory called the Copenhagen interpretation.

A recent experiment called the TEQ partnership could help uncover a boundary between the unusual quantum universe and the usual classical ball and bullet environment of classical physics. Researchers are developing a tool to levitate some silicon dioxide or quartz over the next year to measure the nanometers' size - small but considerably larger than individual particles - that physicists have historically used to illustrate quantum mechanics. How big will an entity be and still be able to show quantum conduct? A tennis ball doesn't behave like an electron - we'll never see a ball enter the left or right field at the same time - but what about a nanoscale quartz slice?

The renewed attempt to decide how matter works at the atomic level sees many alternative theories. One of these is known as the Ghirardi-Rimini-Weber theory, named after the three physicists in the 1980s who formulated it. In this theory, microscopic particles exist simultaneously in many states known as superposition but spontaneously collapse into a single quantum state, contrary to the Copenhagen understanding. According to them, the larger an object is, the less likely it is to overlap, which is why matter considered on a human scale is present only in a single state at a given moment and can be interpreted by classical physics.

In theory, we mentioned earlier, random collapses occur, with a fixed frequency per particle per unit of time.

Indeed, the Copenhagen Principle only collapses when the calculation is done, "because you will need a clear physical criterion for both the calculations and the data, and that's exactly what the theory never contains." The theory addresses this measurement problem by suggesting that collapse is not specific to the measurement method itself, and this is where quantum mechanics stops, and classical mechanics begins. In the TEQ experiment, a small fragment of quartz, one-thousandth the diameter of a human hair, will be assisted by an electric field and trapped in a dark location, and locked with slow to near zero atomic motion.

Scientists will then aim a quartz laser to see if the light's scattering shows signs of the moving target. Silicon dioxide movement may indicate a failure that allows the experiment to demonstrate the three physicists theory's predictions convincingly. Although scientists may not see the predicted signs of failure, the experiment would still allow us to know the universe of the number of particles as it merges with the classical universe of common objects. The results obtained may, however, be a fundamental leap in quantum mechanics.

Some scientists attempt to examine quantum mechanics' basic problems and do not want to admit that this is a scandalous scenario. Compared to the three physicists' theory, alternative theories include MULTIDIMENSIONAL THEORETICAL INTERPRETATION, the notion that any scientific event can occur when particles collapse indefinitely across all possible domains and generate an infinite number of alternative universes. Another alternative, called Bohmian mechanics, named after David Bohm, argues that the probabilities involved in quantum experiments represent only a fraction of our knowledge of a system. In fact, an equation of variables currently disguised by physicists drives the process regardless of whether someone controls it.

However, previous quantum studies' results are still unclear and make it difficult to choose one as the most accurate representation of the truth. TEQ physicists are now developing and testing the experiment device that will be developed within a year. They have the ability to see quantum behavior firsthand and potentially extend the limits of quantum mechanics and explain them by observing quantum behavior.

The results will be comparable to decades of searching for dark matter particles: even if they have not yet been seen, scientists know how particles cannot be gigantic. However, scientists understand that dark matter is present.

CONCLUSION

To conclude this introduction to the fantastic world of quantum physics, we can say that today it is the correct theory conceived in science. It is one of the greatest scientific achievements in history.

It is one of the most successful theories in science. Although it may sometimes appear to have a mystical or esoteric aspect, it is actually a practical branch of physics, which has allowed the development of objects such as the laser, electron microscope, transistor, and superconductor.

The particle/wave duality is not something we can easily understand in our minds.

One of the theories derived from quantum physics is that we can create our own reality simply by thinking of it through what is called the law of attraction.